What if you Could

SNIFF LIKE A SHARK?

EXPLORE THE SUPERPOWERS OF OCEAN ANIMALS

by
Sandra Markle

illustrated by
Howard McWilliam

Scholastic Inc.

For Donna Goff and the children of Beebe Elementary School in Naperville, Illinois

The author would like to thank the following people for sharing their expertise and enthusiasm: Professor John Davenport, University College Cork, Cork, Ireland; Dr. Tory Hendry, Cornell University, Ithaca, New York; Dr. Miriam Sharpe, University of Otago, Dunedin, New Zealand; Dr. Shin-ichoro Oka, Okinawa Churashima Foundation Research Center, Okinawa, Japan.
A special thank-you to Skip Jeffery for his support during the creative process.

Photos ©: cover, 1 wood sign: Dimdimich/Dreamstime; cover top right: Paulo Oliveira/Alamy Stock Photo; cover left: Catmando/Shutterstock; cover bottom right: Schnapps2012/Shutterstock; cover background: Sergey Nivens/Shutterstock; cover background: Rich Carey/Shutterstock; cover texture: Ryan Yee/Shutterstock; 4: wildestanimal/Shutterstock; 4 globe and throughout: MicroOne/Shutterstock; 6–7: wildestanimal/Shutterstock; 6: Nature Picture Library/Alamy Stock Photo; 7: Sergey Uryadnikov/Shutterstock; 8: Linda20july/iStockphoto; 10–11: Andrey Nekrasov/Alamy Stock Photo; 10: David Nunuk/age fotostock; 11: Gilberto Villasana/Shutterstock; 12: Auscape/age fotostock; 14–15: Visual&Written SL/Alamy Stock Photo; 14: creativestudioimage/Shutterstock; 15: Norbert Wu/Superstock, Inc.; 16: Stephen Belcher/age fotostock; 18–19: viper345/Shutterstock; 18: ChameleonsEye/Shutterstock; 19: Geoff Marshall/Alamy Stock Photo; 20: Rich Carey/Shutterstock; 22–23: Cigdem Sean Cooper/Shutterstock; 22: Vkilikov/Dreamstime; 23: ADAM Butler/Alamy Stock Photo; 24: Solvin Zankl/NPL/Minden Pictures; 26–27: Solvin Zankl/NPL/Minden Pictures; 26: David Liittschwager/Getty Images; 27: Solvin Zankl/NPL/Minden Pictures; 28: Brian J. Skerry/Getty Images; 30–31: Schnapps2012/Shutterstock; 30: DavidTLC/iStockphoto; 31: Jurie Maree/Shutterstock; 32: Dante Fenolio/Science Source; 34–35: Bluegreen Pictures/Alamy Stock Photo; 34: Solvin Zankl/Alamy Stock Photo; 35: Nature Picture Library/Alamy Stock Photo; 36: Anthony Pierce/Alamy Stock Photo; 38–39: Paulo Oliveira/Alamy Stock Photo; 38: Kyoko Uchida/Alamy Stock Photo; 39: Anthony Pierce/Alamy Stock Photo.

All endangered status listings are based on the International Union for Conservation of Nature (IUCN) Red List of Threatened Species.

Text copyright © 2020 by Sandra Markle
Illustrations copyright © 2020 by Howard McWilliam

Library of Congress Cataloging-in-Publication Data available

ISBN 978-1-338-35607-6 (paperback) / 978-1-338-35608-3 (hardcover)

10 9 8 7 6 5 4 3 2 1   20 21 22 23 24

Printed in Malaysia 108
First edition, June 2020

Book design by Kirk Benshoff and Sunny Lee
Photo Research by Marybeth Kavanagh

What if one day when you woke up,
you found out that overnight you gained an ocean animal's SUPERPOWER?
What if you could shape-shift like a giant Pacific octopus?
Sting like an Australian box jellyfish?
Or have some other ocean animal's cool natural ability?
How in the world would an ocean animal's superpower change your life!?

WHAT IF YOU COULD SNIFF LIKE A GREAT WHITE SHARK?

WHERE IN THE WORLD?

Great white sharks are most common in oceans where the water is between 54°F and 75°F.

A great white shark doesn't sniff the way you do. It swims with its snout straight ahead, forcing water into its nostrils and through short tubes. Before the water flows out again, it passes folds of skin packed with scent-sensing cells. These send messages to the shark's brain. That's how the shark *smells* what's in the water. Wounded animals are easier to hunt, so a great white is always sniffing for blood. It can smell even a tiny amount up to three miles away!

WHAT YOU SHOULD KNOW

Adult Size
As long as 21 feet and weighs up to 2,400 pounds

Life Span
Around 70 years

Diet
Mainly fish, seals, and dolphins

dorsal fin ——

caudal fin ——

pectoral fin ——

GROWING UP

A baby great white shark is called a pup. It develops inside an egg in its mother's body. At first, the pup gets the food it needs from its egg's yolk. When it leaves its egg, it continues to grow inside its mother for another 12 to 18 months. During this time, it eats undeveloped eggs—even some of its siblings! At birth, it's already five feet long, weighs about 75 pounds, and has a mouthful of sharp teeth.

gill slits
Water enters through the shark's mouth, then flows through its gills and out the gill slits. That is how it gets the oxygen it needs from the water.

WONDER WHY?
a great white shark's body has a tall dorsal fin?

Without it, each strong, side-to-side sweep of the shark's caudal fin would roll it over. The dorsal fin keeps a swimming great white's body upright.

nostril

teeth
There are about 300 total. Two front rows are ready to bite, and the rest replace teeth as they're lost.

SUPERCHARGED!

If you could sniff like a great white shark, you could pinpoint smoke in time to prevent forest fires!

GIANT PACIFIC OCTOPUS

WHAT IF YOU COULD SHAPE-SHIFT LIKE A GIANT PACIFIC OCTOPUS?

WHERE IN THE WORLD?

The giant Pacific octopus is most commonly found in shallow water.

A giant Pacific octopus's body is boneless. The only hard parts are its parrotlike beak around its mouth and two platelike shells inside its head where muscles attach. The rest of its big body is like a muscular balloon with flexible arms. So a giant Pacific octopus can fit through anything bigger than its largest hard body part—its beak. It can even squeeze through an opening as small as a soda pop can!

IF YOU COULD SHAPE-SHIFT LIKE A GIANT PACIFIC OCTOPUS, YOU'D NEVER NEED A GATE TO SLIP THROUGH A FENCE.

WHAT YOU SHOULD KNOW

Adult Size
The main body, called the mantle, is about 24 inches long, but it can be as big as 30 feet across between fully spread arms. It weighs up to 600 pounds.

Life Span
About four years

Diet
Mainly clams, abalone, scallops, and fish

mantle

eye

siphon

GROWING UP

A giant Pacific octopus baby is called a hatchling. It develops inside an egg as tiny as a grain of rice. Its mother lays about 100,000 eggs and sticks them along the walls of her den. She guards her eggs until the young hatch about seven months later. Only about as big as a garden pea, each hatchling swims to the surface. It floats there for about three months, eating what it can catch and growing bigger. Then it sinks back to the ocean floor to continue feeding and growing.

WONDER WHY?
a giant Pacific octopus has a large siphon?

It's because that siphon is the body part the octopus uses to jet away from danger. The rest of the time, the siphon is simply part of the octopus's breathing process. Water enters through an opening in the mantle and passes its gills, which extract oxygen. Then the water flows out through the siphon—except when the octopus needs to move fast! Then its muscular mantle squeezes and squirts water out the siphon. At the same time, the octopus can release ink into the squirting water. That creates a dark cloud that hides the octopus while it makes its getaway.

arms
Each of the eight arms is covered with over 200 suckers to grab and hold prey.

SUPERCHARGED!

If you could shape-shift like a giant Pacific octopus, you could easily squeeze onto a crowded sofa.

WHAT IF YOU COULD STING LIKE AN AUSTRALIAN BOX JELLYFISH?

WHERE IN THE WORLD?

Australian box jellyfish are mainly found in warm surface waters close to coasts.

An Australian box jellyfish has as many as 60 tentacles—stringy, flexible parts—hanging from its box-shaped body. These tentacles can stretch out as far as 10 feet. Each one is covered with around 5,000 stinging cells. If the tentacles are touched, or even just brushed against, these stinging cells uncoil—*ZAP*! Wherever one makes contact, it stabs, injecting venom that's deadlier than a cobra's. Even a tiny amount causes extreme pain.

WHAT YOU SHOULD KNOW

Adult Size
Its body is up to 10 inches across and weighs up to four pounds.

Life Span
About one year

Diet
Mainly fish, shrimp, krill, and arrow worms

bell
This jellyfish has a box-shaped, elastic body.

mouth

eye

GROWING UP

A baby Australian box jellyfish has different names at different stages while it's growing up. Females release their eggs into the ocean. When an egg hatches, the baby is called a larva. Soon, the larva settles to the seafloor and attaches to coral or a rock. Then the larva changes shape to look just like a tiny sea anemone. During this stage, it's called a polyp, and it feeds on plankton, tiny living things in the water. After a few months, the polyp goes through a special type of reproduction, called budding. This produces lots of adult jellyfish. These separate from the polyp and swim away.

tentacles

WONDER WHY?
an Australian box jellyfish has so many eyes?

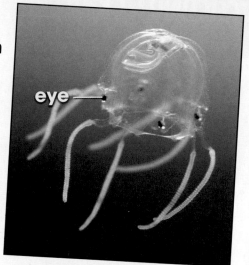

eye

The jellyfish doesn't see the way people do, but it has lots of eyes—clusters of six along the lower edge of each side of its boxy body! These eyes are simply to sense light from dark. They let the jellyfish *see* all the way around itself, so it can avoid floating debris or big predators, such as sunfish. It can also *spy* when small prey comes close enough to sting and eat.

SUPERCHARGED!

If you could sting like an Australian box jellyfish, you'd be a quarterback who *never* gets sacked!

WHAT IF YOU COULD SNAP LIKE A COCONUT CRAB?

WHERE IN THE WORLD?

Coconut crabs are found on islands in the Pacific and Indian Oceans. Christmas Island hosts the world's largest population.

The snapping force of an adult coconut crab's massive claws is the same as a lion's bite! Scientists discovered this using special equipment that compares size to strength. With such force, it's no wonder the coconut crab is able to feast on hard foods, like coconuts. This crab also snaps its powerful claws to warn other crabs to stay away and to defend itself against predators.

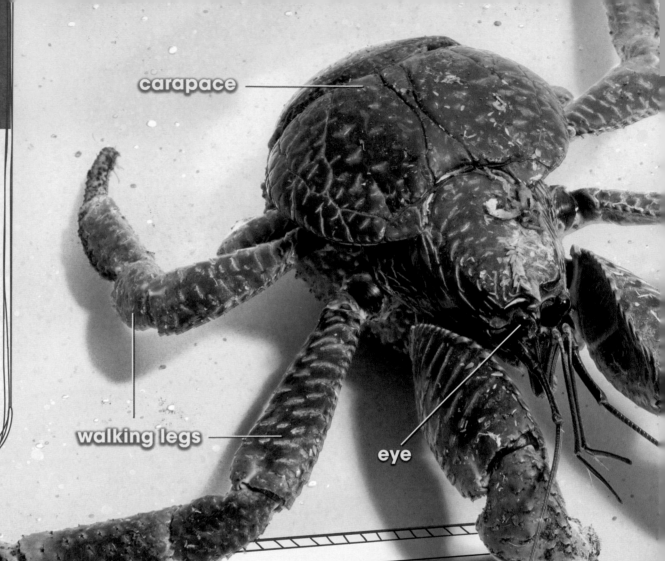

Adult Size
Up to 16 inches long and weighing up to 10 pounds, they are the world's biggest crustaceans—cousins of lobsters and shrimp.

Life Span
As long as 50 years

Diet
Mainly fruit and nuts (including coconuts)

carapace

walking legs

eye

GROWING UP

A baby coconut crab is called a larva. It develops inside a tiny egg stuck to its mother's abdomen. After several months, its mother walks into the ocean in time for her babies to hatch. Each crab larva floats away and feeds on plankton as it continues to develop. After about a month, it becomes a juvenile. Then it has a hard body cover, called a carapace, covering the front of its body but not its abdomen. So it finds a shell on the seafloor and pushes its abdomen inside. As the juvenile eats and grows bigger, it keeps switching shells for one that is larger. Finally, the coconut crab develops its own hard abdominal covering. It's now an adult coconut crab and moves ashore to live on land.

WONDER WHY?

once a year, a coconut crab hides for six to eight weeks?

Hiding is how the crab protects itself while it is shedding its carapace to grow bigger. A new carapace is already under the old one, but it is soft at first. That lets this covering stretch to fit the crab's bigger body. The crab hides out until its new coat becomes armor-hard.

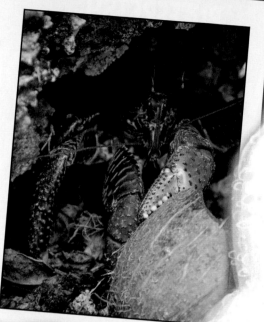

left claw
The left claw is always larger than the right.

antenna

SUPERCHARGED!

If you could snap like a coconut crab, you'd easily lead expeditions through extra-thick jungles.

STARRY PUFFERFISH

WHAT IF YOU COULD PUFF UP LIKE A STARRY PUFFERFISH?

WHERE IN THE WORLD?

Starry pufferfish are most often found around coral reefs.

A starry pufferfish can easily expand to two or three times its normal size. The pufferfish puffs up by quickly gulping a lot of water into its elastic stomach. Its skin, which is made of wavy folds, also stretches. So when a hungry predator like a tiger shark swims close, the fish puffs into a ball that's hopefully too big to bite. Getting bigger fast makes a good defense!

IF YOU COULD PUFF UP LIKE A STARRY PUFFERFISH, YOU'D BE THE MAIN ATTRACTION IN EVERY PARADE!

Adult Size
About 21 inches long and weighs around nine pounds

Life Span
Up to eight years

Diet
Mainly sea urchins, sponges, crabs, and coral

dorsal fin

caudal fin

pectoral fin

anal fin

GROWING UP

A baby pufferfish is called a fry. A female releases three to seven eggs in the warmer water close to shore and swims away. The eggs float for about a week. Then the fry hatch, but they aren't fully developed yet. A fry's body has a hard shell that protects it for the few extra days needed for its fins to grow. Once that happens, the shell cracks and falls off.

skin
Instead of its skin being covered with scales like most fish, it has a poisonous, snot-like coating.

eye

mouth

WONDER WHY?
a starry pufferfish's eyes aren't always looking in the same direction?

That's because each eye can turn separately. So a starry pufferfish can look in two different directions at once to watch for prey and predators.

SUPERCHARGED!

If you could puff up like a starry pufferfish, you'd save your friends a spot in the bleachers without breaking a sweat!

WHAT IF YOU COULD LIGHT UP LIKE A FIREFLY SQUID?

WHERE IN THE WORLD?

Firefly squid live along coastal shelves in the ocean's twilight zone, where only faint light filters down.

Firefly squid's bodies are covered with special light-producing organs. Each organ produces a chemical called luciferin and a protein called luciferase. The reaction between the chemical and protein when they are combined with oxygen from the water gives off cool-blue light. These lights can be flashed all at once or in patterns to attract a mate. The light also lures prey close enough for the firefly squid to catch with its tentacles.

IF YOU COULD LIGHT UP LIKE A FIREFLY SQUID, YOU'D BE THE SPY EVERYONE COUNTS ON TO SEND SECRET MESSAGES.

Adult Size
About three inches long and weighs only about a third of an ounce

Life Span
About one year

Diet
Mainly small fish

fin

mantle

GROWING UP

A baby firefly squid is called a paralarva. Every year from March through June, thousands of adult firefly squid gather near the ocean surface at mating sites. Each female releases as many as 20,000 eggs in jellylike strings that can be up to four feet long. The eggs hatch in 6 to 14 days. Each firefly squid paralarva looks just like an adult, only much smaller. It spends its days in the dark depths. But it comes to the surface at night to feed on prey and grow bigger.

eye

siphon
This is where water is ejected to make the firefly squid jet away from danger.

WONDER WHY?
a firefly squid's tentacles are longer than its arms?

The tentacles are long enough to stretch out into the dark ocean. The light-producing organs at the tentacle tips lure fish prey close. The tentacles also have suckers that are used to snag the prey. Then it pulls the prey close enough for the arms to grab and move to its mouth.

arm

SUPERCHARGED!

If you could light up like a firefly squid, you'd be the school's star crossing guard.

LOGGERHEAD SEA TURTLE

WHAT IF YOU COULD ARMOR UP LIKE A LOGGERHEAD SEA TURTLE?

WHERE IN THE WORLD?

Loggerhead sea turtles are found in every ocean except the cold Arctic and Antarctic.

A loggerhead sea turtle's shell is its body's armor. And it's hard enough to protect it from being chomped by a shark! The turtle's shell is formed from about 60 bones, which are part of its rib cage and backbones. These bones are covered with hard, overlapping plates formed from keratin, the same material that makes up human fingernails. Because the main part of its shell is made of its bones, as the turtle's body grows bigger, so does its armor.

IF YOU COULD ARMOR UP LIKE A LOGGERHEAD SEA TURTLE, YOU'D NEVER WORRY ABOUT SLIPPING ON AN ICY SIDEWALK.

WHAT YOU SHOULD KNOW

It's endangered!

Adult Size
About three feet long and weighs as much as 250 pounds

Life Span
May be more than 60 years

Diet
Mainly horseshoe crabs, clams, and sea urchins

eye

head
Unlike some kinds of turtles, it can't pull its head into its shell to stay safe.

beak

front flipper

GROWING UP

A baby loggerhead sea turtle is called a hatchling. Each baby develops inside a leathery-shelled egg. A female produces the eggs and buries them in a nest she digs on a sandy beach. One nest may contain as many as 300 eggs. How warm the egg stays determines the hatchling's sex. Warmer eggs produce females, and cooler eggs produce males. The eggs hatch after about 60 days. Then, each hatchling crawls to the water and swims away. It is on its own to grow up in the ocean.

upper shell

WONDER WHY?
a loggerhead sea turtle has such big flippers?

The turtle uses its flat, broad, winglike front flippers to propel itself through the water. The back flippers act as rudders for steering. Females also use their back flippers to dig a nest hole in the sand for their eggs.

— back flipper

bottom shell

SUPERCHARGED!

If you could armor up like a loggerhead sea turtle, you'd be the best movie stuntperson in Hollywood!

WHAT IF YOU COULD ENCHANT LIKE A FEMALE DEEP-SEA ANGLERFISH?

WHERE IN THE WORLD?

Deep-sea anglerfish float along in the ocean depths. Scientists are continuing to investigate its range.

Only the female deep-sea anglerfish has a lure to enchant prey. It is formed by a stiff, supporting ray in the fin on top of her back. There are tiny, glowing bacteria living in a bulb-shaped part at the tip of the lure. That glowing part stretches out just far enough to dangle in front of the fish's big mouth. In the dark ocean depths, the female anglerfish only has to stay still to hide. Then she wiggles her light-tipped lure. Any little fish enchanted enough to come check it out is dinner. *GULP*!

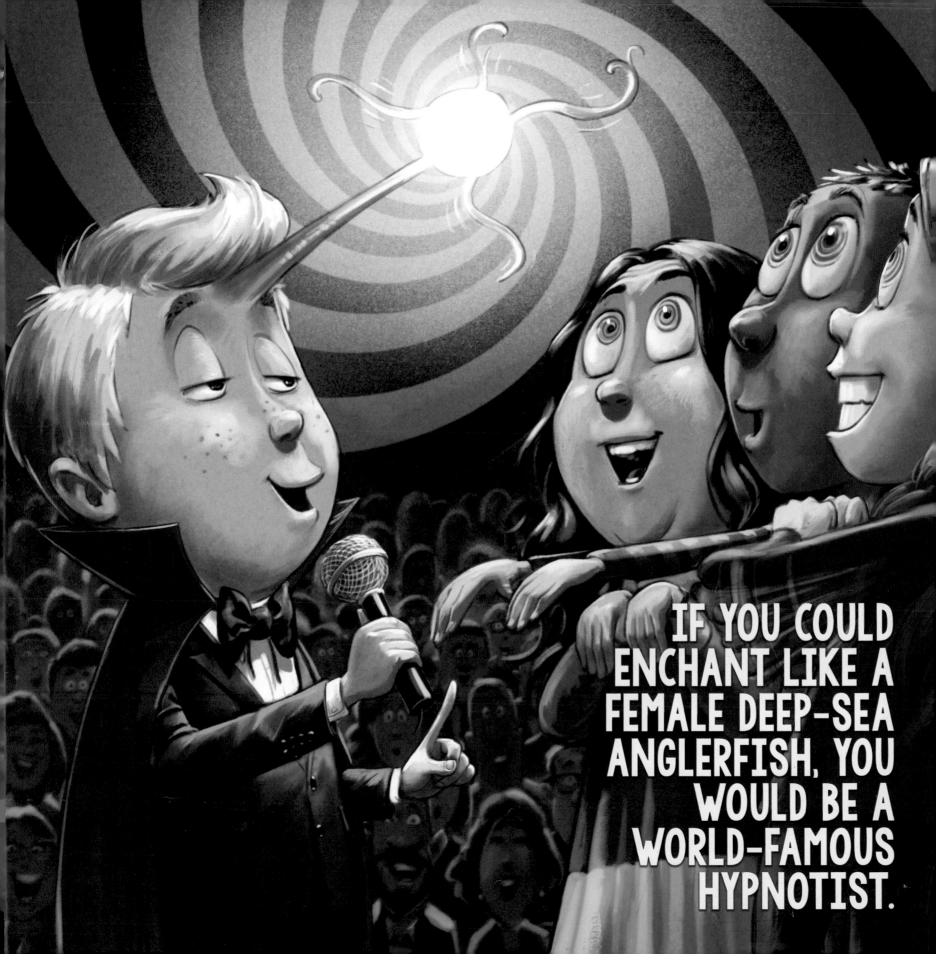

Adult Size
Up to 30 inches long and weighs as much as 110 pounds

Life Span
Around 30 years

Diet
Any fish they can catch!

lure

eye

mouth

GROWING UP

A baby deep-sea anglerfish is called a larva. A female releases as many as a million tiny eggs at a time. The eggs are within a huge, thin, gelatin-like sheet that floats to the ocean's surface. There, the larvae hatch. As each larva feeds and grows bigger, it lives at deeper and deeper levels—down to the ocean depths. Males attach to females, stay tiny, and are carried along with them. Only the females develop lures.

body
Its basketball-shaped body floats easily.

skin

WONDER WHY?
a female deep-sea anglerfish has such a big mouth?

Prey is hard to come by deep down in the ocean. So an anglerfish needs to be able to take advantage of whatever fish it can attract close enough to catch. That's why an anglerfish needs to have a big mouth it can open really wide. And its stomach can also stretch to hold a *big* meal.

SUPERCHARGED!

If you could enchant like a female deep-sea anglerfish, you'd be the most spellbinding storyteller on the camping trip.

WHAT IF YOU COULD GLIDE LIKE A BLUE FLYING FISH?

WHERE IN THE WORLD?

Blue flying fish are found around the world in tropical ocean water.

When a blue flying fish leaps into the air, its winglike fins move forward and sideways, like a person spreading their arms wide. These fins lock into place, with lots of stiff rays for support. Once in the air, its fins let a blue flying fish glide as far as 328 feet in one leap! With any luck, that's far enough to escape a fast, hungry predator, such as a tarpon or a swordfish.

IF YOU COULD GLIDE LIKE A BLUE FLYING FISH, EVEN CATCHING THE SCHOOL BUS COULD BE AN ADVENTURE!

Adult Size
About 8 inches long and weighs four ounces

Life Span
About five years

Diet
Mainly zooplankton, tiny animals living near the ocean's surface

caudal fin ————

GROWING UP

A baby blue flying fish is called a larva when it hatches and a fry when it's older. Females release as many as 300 eggs at a time. These eggs have many tiny, sticky strings that anchor them to seaweed floating in the water. Each larva hatches in about a week. But it's attached to the yolk sac that was in its egg. It still needs this for food while it develops a little more. By the time the yolk is used up, the larva can feed itself. Then it's a fry. Still little, it hides among other tiny animals floating at the surface while it feeds and continues to grow bigger.

eggs

pectoral fin

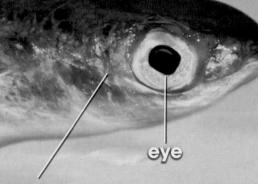

eye

gill flap
Covers and protects gills.

WONDER WHY?
a blue flying fish's forked tail is much longer on the bottom?

During its glide, the fish eventually sinks back toward the water. The longer section of its caudal fin touches the ocean's surface first. When it beats fast against the water, this longer section launches the fish up into the air again, so it can glide even farther.

SUPERCHARGED!

If you could glide like a blue flying fish, you'd set an unbreakable world long jump record!

HOME SWEET HABITAT

All living things need a habitat to call home. This special place supplies the oxygen, water, food, shelter, and space needed to live and produce babies. Earth offers a number of very different habitats. One of these is its oceans.

Ocean habitats are divided into two main areas: **coastal** (close to shore) and **open ocean** (all the rest). From the top down, the open ocean is also divided into different habitat zones.

SUNLIGHT ZONE
(down to about 660 feet)

Most ocean plants and animals live in this sunlit, warm water.

TWILIGHT ZONE
(from 660 feet down to 3,300 feet)

There is only faint light in this cooler water zone. Fewer animals live here, and there are no plants.

MIDNIGHT ZONE
(from 3,300 feet down to 13,000 feet)

This zone is cold and dark with high water pressure. The only light is from animals that produce their own light.

ABYSS ZONE
(13,000 feet down to seafloor)

The water here is the coldest and darkest in the ocean. The water pressure is very high. But some animals still live here, and some others visit.